Waste Plastics 2020

Topical Conference at the 2020 AIChE Annual Meeting

Online
16 – 20 November 2020

ISBN: 978-1-7138-2295-0

Printed from e-media with permission by:

Curran Associates, Inc.
57 Morehouse Lane
Red Hook, NY 12571

Some format issues inherent in the e-media version may also appear in this print version.

Printed with permission by Curran Associates, Inc. (2021)

For permission requests, please contact AIChE
at the address below.

AIChE
120 Wall Street, FL 23
New York, NY 10005-4020

Phone: (800) 242-4363
Fax: (203) 775-5177

www.aiche.org

Additional copies of this publication are available from:

Curran Associates, Inc.
57 Morehouse Lane
Red Hook, NY 12571 USA
Phone: 845-758-0400
Fax: 845-758-2633
Email: curran@proceedings.com
Web: www.proceedings.com

TABLE OF CONTENTS

600539 Quantitative Risk Assessment of Solvothermal Conversion of Waste Plastics

Abstract:

Solvothermal conversion is a thermochemical process, where biomass is treated in the presence of a solvent at a temperature higher than the boiling temperature of this solvent. In the solvothermal process, the solvents play a significant role in overall conversion, distribution of products, and bio-oil yield and properties. For instance, the presence of organic solvents during the solvothermal process allows the product to contain important functional groups. At the same time, most of the organic solvents are flammable in nature. In addition, the solvothermal process is generally a high temperature and high-pressure process. So, a significant risk should always exist associated with the solvothermal process.

Due to the success in a laboratory scale, people are thinking about it's commercialization. As soon as we think about the large scale, the risk associated with the design comes to the place. As part of the conceptual design, the hazards associated with the solvothermal process should be reviewed and identified the critical assumptions to support safe operations. To the best of authors' knowledge, no study has been reported on the risk associated with the solvothermal process and how it changes with the plant capacity. Therefore, a detailed safety study of the solvothermal process is urgently required.

In this study, solvothermal processes, the capacity of handling 1 to 100 Mt/day waste plastic as feedstock, has been considered. A complete hazard identification along with the risk assessment were completed. During the risk analysis, two different release cases were considered i) loss of containment and ii) leak from the reactor. In this analysis, different possible ways to propagate the release scenarios such as flammable and toxic gases, Boiling Liquid Expanding Vapor Explosions (BLEVE), Jet fires, etc. were studied. Results showed that the risk associated with the process is not linearly proportional to scale-up of the plant capacity.

611520 Global Plastic Waste Trade: Assessing Potential Interventions for Improving Environmental Sustainability

Abstract:

The amount of plastic waste generated globally has increased significantly over the past few decades, with sustainable end-of -life management remaining prohibitively expensive in many regions. As a result of this, many countries export some of their plastic waste, with developing countries most often taking in the waste from developed countries. Due to the present disparities in waste management capacity between trading countries, this trade network resulted in over 2 billion kilograms of extra mismanaged plastic waste in 2010. Using publicly available data, we create a network model of the global plastic waste trade. Additionally, we construct scenarios to quantify the potential impacts of a range of interventions on the plastic waste trade network. These models are used to evaluate the effectiveness and efficiency of these interventions as options to steer the network to a more desirable trajectory, where there is a net negative quantity of mismanaged plastic waste. Preliminary results suggest that import bans are not effective, as the displaced flows are generally redirected to other developing nations with similar poor waste mismanagement rates. However, strategies that aim to increase waste management capacity while decreasing its cost in key developed countries may be effective at redirecting plastic waste towards more environmentally friendly end-of -life scenarios. Such strategies could have a dual effect of both decreasing global plastic waste mismanagement rates and aiding in a transition towards a more circular economy of plastics.

610270 Overview and Assessment of Waste Plastics in the Ocean Environment

Abstract:

Humans are currently contributing to a major decline in the world's biological diversity. The problem is so critical that all human impacts combined might accelerate the present extinction rates from 1000 to 10,000 times the natural rate. The threat to marine life in the oceans comes in various forms such as dumping of waste, overexploitation and harvesting, pollution, alien species, land reclamation, dredging, and global climate change. One particular type of human impact that represents a major threat to marine life is the pollution caused by plastic debris. Plastic has become one of the most common forms of pollution in the ocean and is causing problems to become more and more severe each day, impacting both the natural environment and human life.

Since the end of World War II, plastics have been produced in large and exponential quantities. They represent a diverse group of materials which are derived from petrochemicals. Their production has increased exponentially from 1,700,000 tons in 1950 to more than 280,000,000 tons in 2011. Although plastics are recyclable, they contribute to producing a significant amount of waste globally. For example, from 2010 to 2011, Australia produced 1,433,046 tons of plastics among which only 20% was recycled. Also, 37 percent of that plastic was for the single-use disposable packaging. Plastics break down slowly through a combination of oxidation, photodegradation, and mechanical abrasion. Thick plastics survive for long times even if they are subjected to direct sunlight. When plastics are submerged under water or in sediments, protected from UV radiation, they can survive for an even longer time. Although biodegradable plastics have been developed as a solution, it has not had a long-term effect yet. Plastics have a lower proportion of biodegradable materials and thus microscopic plastic fragments are left behind.

The existence of widespread plastics in the ocean was not observed until 1970. As a result, plastic pollution of oceanic waters was not a matter of concern prior to that time. The production of plastics that are not biodegradable and the amount of waste have been increasing worldwide, causing increased levels of contamination and posing an escalating threat to the marine environment. As the ocean is downstream from almost every terrestrial location, it receives most of the plastic waste generated on land. Every year, it receives several million tons of debris and much of this is discarded plastic litter.

Plastic pollution has an adverse impact on our oceans and the health of wildlife. It is causing harm to the creatures that live in the ocean, from coral reefs to turtles gagging on straws, to whales and seabirds. These creatures may starve because their bellies become so jammed with bits of plastic that there's no room for substantial food. Another recent concern regarding plastic pollution is the impact on food webs and marine ecosystems. The concentrated toxins of plastic materials may be delivered to animals via ingestion and transferred to their food webs. At least 26 species of cetaceans have been documented to ingest plastic debris.

This paper examines reported evidence and effects of plastic pollution in oceans and the marine environment.

611055 Recent Advances in Plastics Upcycling

Abstract:

This talk will review our group's recent efforts in the deconstruction and upcycling of today's plastics using chemical catalysis, enzymes, and engineered microbes. Examples include depolymerization of PET using enzymatic catalysts, depolymerization of polyolefins with chemical catalysis, and upcycling of plastics-derived intermediates using biological funneling. Analysis of the major plastics consumed in the United States will also be presented, describing the supply chain energy and greenhouse gas emissions associated with commodity plastics as a foundational dataset for evaluating new circular economy and bio-based plastics technologies.

611770 Utilization of Waste Plastics As Feedstock for the Production of Useful Chemicals Using an Integrated Approach

Abstract:

Plastic is a non-biodegradable substance because its chemical configuration is resistant to environmental degradation thus resulting in environmental pollution. Plastic is essentially an indispensable commodity, relatively cheap and its production continues to rise because of its numerous industrial and domestic applications. The accumulation of waste plastics continues to grow as its usage increases around the globe. Over 350 million tons plastics are produced annually and over 8 million tons of waste and discarded plastics end up in the ocean. It has been projected that if the growing rate continues in this trend, the number of micro-plastics in the ocean would outnumber the current number of fishes in the marine ocean by 2050. This paper focuses on an integrated approach to utilization of waste plastics as feedstock for the production of useful chemicals. Though pyrolysis is gaining attention in the conversion of waste plastics into fuel, yet most studies in this area are still in the laboratory scale and with limitations. In this study, different categories of waste plastics are subjected to thermochemical treatments integrated with catalytic pyrolysis. Technologies used for the thermochemical treatment are discussed and the thermochemical products are made to undergo distillation and separation processes to analyse, characterize and evaluate the properties. Conversion of waste plastics into valuable chemical products will significantly help to solve global environmental issues associated with plastic wastes.

599033 Study of Recycled Carbon Fiber Length Distribution Measurement and the Effect to the Tensile Properties of Pulp-Molded CFRP

Abstract:

Carbon fiber reinforced plastics "CFRP" are performed as a promising solution of light weight demand for energy saving of transportations. Especially in aerospace industry one model of aircraft includes more than 50% weights of CFRP parts. The demands of CFRP are increasing due to the expansion of the application for automobiles, pressure vessels and the other markets. On the other hands, the poor recyclability of CFRP is one of big issue in environmentally and economically, motivates to work with recycle study in many institutes and companies. We had developed a CFRP recycle system through two stage thermal treatments including pyrolysis of plastics in inert conditions and oxidation of residue of carbonized plastics. The recycle carbon fiber obtained by the system indicates the physical properties of more than 80% tensile strength comparison to virgin carbon fiber and less than 1% residual carbonized plastics in a usual treatment condition. From the point of the recycle carbon fiber users view, it is important not only the tensile properties and impurities but also the fiber length and the distribution. In this report, we studied the methodology for a quality measurement of recycle carbon fiber length , and the effect to tensile properties of a pulp-molded CFRPs.

For the efficient measurement of fiber length, a commercially available optical scanner was introduced to the measurement procedure. To be homogenized the sample carbon fiber, the chopped carbon fiber was dispersed into water solution of a surfactant reagent. A part of the dispersion liquid was put into a pair of thin clear plastic film to prepare the scanning sample. The fiber length was measured by a personal computer with a software from the scanned image file. Both of virgin and recycle carbon fiber were applicable for the measurement procedure. The results of virgin carbon fiber were nearly equal to cut length, support the measurement value are reliable. In the case of recycle carbon fiber, the results were broader distribution including at least two different major group. One of major group value was equal to the cut length, the other group value was roughly 1.5 times greater than the cut length.

A pulp molding procedure was selected to convert the recycle carbon fiber to CFRP due to the easiness of machine preparation and the smallness experimental scale. A carbon paper was prepared as an intermediate product of CFRP. The carbon fiber was dispersed into a water solution of surfactant reagent and binding reagent, and it was filtered off by a metallic mesh to obtain the carbon paper. A polypropylene film was selected as a representative of thermoplastics, and an epoxy resin was selected as a representative of thermoset plastics.In both cases, the value of tensile test showed best result in CFRP using 10 mm carbon fibers.

593126 Recycling Plastic Waste into High-Valued Engineering Aerogel Materials

Abstract:

In 2017, Singapore generated 7.7 million tonnes of waste, a sevenfold increase from 40 years ago and enough to fill 15,000 Olympic-size swimming pools. Plastic was the largest category of waste generated but only 4% were recycled, the lowest of all waste categories. This seemingly hard-to-recycle material has been widely featured as a hazard to the environment, polluting oceans and clogging up landfills. Plastic releases toxic pollutants when burnt and can take even thousands of years to break down. Currently, the recycling of plastic waste is mainly for low-value applications and simple uses, which results in a lack of appreciation for the value of this raw material as recyclable waste. The previous lack of possible solutions could also have contributed to public perception that there isn't more they can do to help alleviate the situation.

A new ground breaking technology was created using the freeze-drying process, which makes use of environmental waste to solve environmental problems, by turning plastic waste into novel, high-value aerogel materials. Aerogel is a class of porous material that exhibits an uncanny array of extreme materials properties. Due to its highly insulating and absorbent qualities, the material has a wide range of applications including, but not limited to, heat and sound insulation, high-temperature garments, carbon dioxide and toxic gas filtering, oil spill cleaning and medical devices. At the end of their useful life, the used aerogel can be safely grounded into fine particles and discarded.

This new technology is able to reduce the duration of developing aerogel to less than a day - 20 times faster than the current commercial processes. The method also uses 70% less energy and results in reduced polluting emissions in the air and water. The current project aims to take the research further and help this revolutionary technology gain more international recognition, build awareness and generate interest in how much impact such a novel environmental solution and engineering application will bring about.

This breakthrough could be the solution to the world's plastic waste problem. One of the least recyclable wastes that usually ends up polluting oceans and clogging up landfills could potentially be turned into a life-saving material for fire insulation and cleaning up oil spills. It is important to ensure that such a technology is made publicly aware and translational in facilitating waste management, so that the impact can be further amplified.

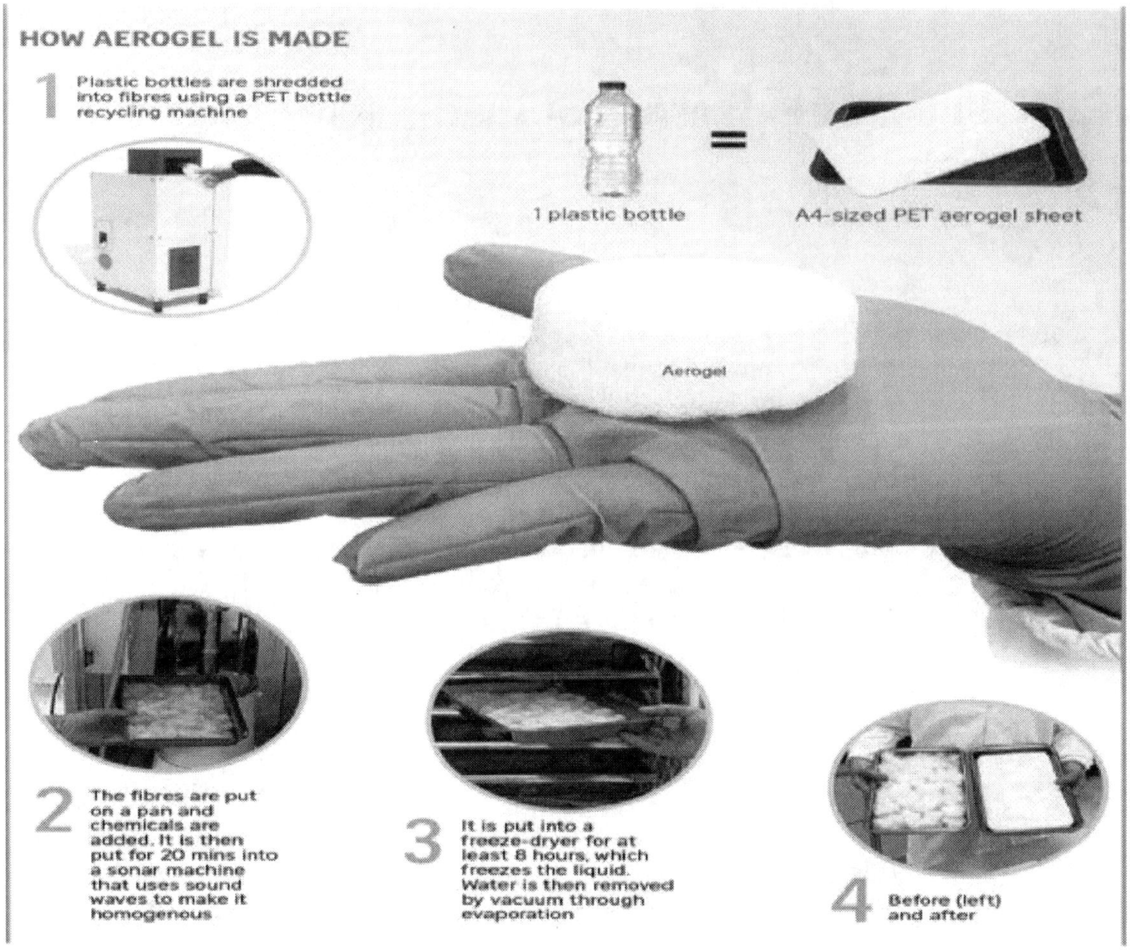
HOW AEROGEL IS MADE
1 Plastic bottles are shredded into fibres using a PET bottle recycling machine
1 plastic bottle
=
A4-sized PET aerogel sheet
Aerogel
2 The fibres are put on a pan and chemicals are added. It is then put for 20 mins into a sonar machine that uses sound waves to make it homogenous
3 It is put into a freeze-dryer for at least 8 hours, which freezes the liquid. Water is then removed by vacuum through evaporation
4 Before (left) and after

607159 Recycling of Multilayer Plastic Packaging Materials By Solvent-Targeted Recovery and Precipitation

Abstract:

Many plastic packaging materials manufactured today are composites made of distinct polymer layers (*i.e.*, multilayer films). Billions of pounds of these multilayer films are produced annually, but no technologies exist to recycle them back into their pure resins without breaking the films down into their original monomers. Herein, we demonstrate a new strategy we call Solvent-Targeted Recovery And Precipitation (STRAP) to deconstruct multilayer films into their constituent resins using a series of solvent washes that are guided by thermodynamic calculations of solvent-polymer solubility. We show that the STRAP process is able to separate three representative polymer resins (polyethylene, ethylene vinyl alcohol, and polyethylene terephthalate) from a post-industrial multilayer film with near 100% material efficiency. In this process a single polymer layer was selectively dissolved in a solvent system to be later separated from the insoluble layers and then precipitated by the addition of an antisolvent and/or decreasing the solvent temperature. The process was repeated for each of the polymer layers in the multilayer film and 100% recovery was obtained for each. This resulted in segregated streams of chemically pure dry polymers that can be much more efficiently recycled. The FTIR spectra of the separated polymer fractions were comparable to the virgin resins and less than 1100-ppm residual solvent was detected with head-space GC-FID. The glass transition temperatures for the PET and EVOH fractions were similar to the virgin materials. The solvent selection and process conditions for STRAP were based on quantifying polymer and solvent thermodynamic properties using three computational methods: Hansen Solubility Parameters (HSPs), molecular dynamics (MD) simulations, and COnductor-like Screening MOdel for Realistic Solvents (COSMO-RS). An economic analysis was performed to assess the feasibility of STRAP at a larger scale. The total capital investment for this technology was estimated to be 13.92 million dollars with a minimum selling price (MSP) of $1.34/kg when the feed rate was assumed to be 3,000 ton/year of multilayer films.

609707 Microplastics Characterization and Impact on Small Fish of the Thermaic Gulf

Abstract:

Microplastics (MPs) are detected worldwide in the marine environment. Plastic production and use have increased steadily over the past 50 years, with global production reaching 335 million tonnes in 2016 (Plastics Europe 2015). Microplastics, because of their small size, can be transferred from the marine environment to organisms, and subsequently pass through the trophic transfer. Ingestion of microplastics has been documented in many seafood species from various places around the world, including fish, bivalves, and mussels (Toussaint et al. 2019). The primary uptake pathways of MPs in the most aquatic organisms are gastrointestinal tract and gills (Su et al. 2019). MP accumulation in aquatic organisms poses an emerging food safety concern. The human health effects depend on the actual exposure concentrations and migration pathways in human body. It is inevitable that humans' exposure to microplastics depends on the levels of seafood consumption worldwide. Therefore, human health risks assessment associated to MPs daily intake represents an important issue for public health. Moreover, the ability of microplastics to accumulate toxic chemical compounds raises a serious concern over the transfer of hazardous compounds to marine animals and subsequently to humans. The aims of this study are (i) to quantify the number of microplastics with diameter range between 100 μm to 5 cm in beach sediments and surface water of Thermaic Gulf, (ii) to identify their spatial distribution and polymeric characterization and (iii) to investigate the presence of MPs in marine organisms. Moreover, the estimation of human daily exposure to MPs due to fish consumption was conducted.

The Thermaic Gulf is affected considerably by urban, industrial and touristic activities. Additionally, the pollution levels can be increased by the re-routing of four rivers to the Thermaic Gulf. The latter is an area constituting the northwest corner of the Aegean Sea. A total of 25 water surface samples were collected at 500 m, 5 km and 10 km distance from the coast during the winter (February - March 2018). Surface samples were collected using neuston net with cod end (300 μm mesh size). For beach sediment sampling, Box Corer grabs were used. In order to ensure representativeness of the samples, 126 samples were collected from five different collection points across the coastline of the Thermaic Gulf. Finally, a total of 75 fish were sampled from Thermaic Gulf. The examined fish species were, *S. colias*, T. trachurus, C. linguatula, B. boops, S. pilchardus, E. encrasicolus, S. maena. The total length and wet weight of fish was recorded. The gills and the gastrointestinal tract of each specimen were separated and weighted. For plastic particle analysis, samples were observed under a stereoscope with a LED light and measured with micrometer ocular lens. Plastic particles were characterized and classified by colour, size and shape. Additionally, the polymer fingerprint was detected using RAMAN spectroscopy, and the bioconcentration factor (BCF) was also estimated. The daily intake dose of MPs concentration in fish was calculated based on the assumption that the ingestion rate is 53 g/day and the body weight is 70 kg for adults and 16 kg for children.

A total of 20000 microplastics have been isolated from all 25 surface water samples, with a range of 190 to 6500 total particles per sample. The average concentration was 1.88 items/m^3 or 750,846 items/km^2, and the highest concentrations are found within a distance of 10km from the shore and in the area of Epanomi. The most predominant color is white and the most frequently found shape is fragment. The majority of microplastics have size between 1.0 - 1.5 mm. Regarding the beach sediments the average concentration was 28.67 items/m^2. Plastic fragments were found to be primarily isolated plastic particles.

Polyethylene (PE) and Polypropylene (PP) were the most abundant chemical species, a fact that was in accordance with literature findings, where these polymers account for the majority of the plastic particles floating in Mediterranean Sea. The examination of seven different fish species showed the negative consequences of the microplastics presence in aquatic ecosystems. In all examined fish, microplastics were found in gills and the gastrointestinal tract. The highest concentration was found in the *C. linguatula* population, 0.45 ± 0.11 items/g, while *Trachurus trachurus* had the lowest one 0.74 ± 0.5 item/g. Therefore, this indicates that the bathypelagic fish are the most affected in comparison with the mesopelagic and the epipelagic fish. In this case, the most predominant color was white and the most predominant shape was fiber. The presence of larger microplastics in fish's organism could suggest their further fragmentation and presence in edible parts of fish. The BCF values were higher in gills than in gastrointestinal tract with average estimates of 695 and 896, respectively, for MPs < 1000 μm, whereas the estimated BCF values for MPs < 100 μm for gills were 22. The estimated daily intake of MPs (< 500 μm) in the wake of fish consumption was 15.7 p/kg/body-weight/day for adults and 9.07 for children.

This study reports the results of a monitoring campaign of microplastics in three different media. These data represent the first attempt to present the current state of microplastics in the Thermaic Gulf and to establish the baseline level for a future study. Given the significance of Thermaic Gulf's port and its morphological features that generate important quantities of waste originating from the South European counties the evaluation of plastics' presence in Thermaic Gulf is mandatory to apply strategies to reduce the quantity of microplastics in its waters. The results showed a higher abundance of microplastics at 5-10 km distance from the coast. Analysis of the size distribution of microplastics in surface samples showed that in the whole dataset the most abundant size class was 0.5 - 1 mm, while for sediment it was 3.5 - 4 mm. Higher concentrations of MPs in GI tract and gills were observed in size class ranging between 500 μm to 1000 μm. The results revealed high values of BCF factor, indicating that the higher absorption occurs through gills. The Estimated Daily Intake shows the quantity of MPs ingested daily by fish consumption. Although there is no standard dose for MP ingestion and information on exact toxicity of different plastic types in the human body, the consumption of high daily doses can threaten the health of consumers.

REFERENCES

Plastics Europe (2015). Plastics – the Facts 2015: an Analysis of European Plastics Production, Demand and Waste Data**:** 30 pp.

Su, L., H. Deng, B. Li, Q. Chen, V. Pettigrove, C. Wu and H. Shi (2019). "The occurrence of microplastic in specific organs in commercially caught fishes from coast and estuary area of east China." J Hazard Mater **365**: 716-724.

Toussaint, B., B. Raffael, A. Angers-Loustau, D. Gilliland, V. Kestens, M. Petrillo, I. M. Rio-Echevarria and G. Van den Eede (2019). "Review of micro- and nanoplastic contamination in the food chain." Food Additives & Contaminants: Part A **36**(5): 639-673.

613796 Critical Issues of Waste Plastic Management in Developing Regions

Abstract:

An overview of critical issues in waste plastic management faced by developing regions

613373 Legal Drivers for Waste Plastic Law and Regulation

Abstract:

Environmental law and policy are designed to mitigate the harmful effects of pollution. Yet, plastic generally tends to fall outside the scope of stringent environmental regulation, depriving entities producing or distributing plastic from specific critical incentives that could best support their development of meaningful sustainable approaches. Familiar approaches to designing environmental law and policy is not specifically designed to address persistent synthetic materials such as the current suite of synthetic commercial plastics distributed in the market today. Environmental programs generally focus on addressing pollution by mitigating exposure to chemicals that pose a threat of acute and chronic toxicity including carcinogenicity. The synthetic macro and microplastics in our environment, while likely leaching some chemical pollutants, have observably caused harm through their persistence in the environment resulting from their inert nature. Through their characteristics of fracturing into smaller and smaller particles while not dissolving, reacting or degrading, synthetic plastics may result in significant acute and chronic harm from physical obstruction of biological processes, whether as macro, micro or nanoparticles. This harm is generally not mitigated through conventional environmental regulation focusing on mitigating chemical toxicity. Ongoing research that better defines this harm, and provides reproducible parameters, metrics and nomenclature for such harm, will support the development of new approaches to mitigating such harm. These new approaches can support better application of existing law and policy, while also informing development of more tailored authority to address this type of environmental pollutant. These new approaches also support litigation and other methods of distributing the burden of risk and harm better defined from this research once the harm can be quantified. This presentation will summarize the current status of global policy approaches, including the United States, and how both ongoing efforts to qualify and quantify harm, and these new legal and policy measures, serve to provide significant incentive for all plastic dependent industry sectors to seek more aggressive approaches to keeping plastic out of the environment.

613246 Public Health Perspectives Associated with Waste Plastics

Abstract:

Each stage of the plastic lifecycle poses major risks to human health. Worldwide, most people are exposed to plastics at all stages of their lifecycle. Plastic pollution is pervasive worldwide infiltrating our environment, wildlife and marine animals, our food supply, and our human health. Health problems associated with waste plastics include: various forms of cancer; diabetes; several organ malfunctions; impacts on skin, eyes, and other sensory organs; birth defects, among other impacts. Additionally, there are harmful impacts associated with plastic additives, including bisphenol-A, heavy metals (e.g., cadmium and lead); flame retardants, phthalates, perfluorinated substances (PFAS), endocrine disrupting compounds, and other chemicals Plastic production has increased from 2×10^{6} metric tons in 1950 to 3.80×10^{8} metric tons in 2015.

Direct environmental exposure to plastics is a result of extraction and transport, refining and manufacture, consumer products and packaging, waste management, fragmentation and micro-plastics formation, cascading exposure as plastics degrade, and ongoing environmental exposures. Humans are exposed to microplastics and associated toxic chemicals through inhalation, ingestion, and direct skin contact throughout the life cycle of plastics. Public health is exacerbated by the fact that more than 2/3 of all plastic produced remains in the environment, as debris in the oceans and waterways, solid waste and litter on land, micro- or nano-particles in the air or agricultural soils, microfibers in water supplies, or microparticles in the human body, etc. As a result, plastics have permeated our air, our soil, our water, and our bodies.

This plenary presentation addresses public health perspectives associated with waste plastics, and offers approaches for minimizing waste plastic formation.

602978 Bio-Based Benzoxazine Monomers and Polymers Based on Difuran Diamine (Dfda)

Abstract:

A new class of bio-based benzoxazine monomers based on difuran diamine (DFDA) obtained by the coupling of furfuryl amine using formaldehyde will be discussed. The benzoxazine monomers were synthesized by reacting DFDA with a number of bio-based phenolic compounds and formaldehyde. These systems were purified by precipitation. Depending on composition, the resulting solid powders melt at temperatures ranging between 70°C and 150°C to form low viscosity liquids that can be readily used to impregnate fiber reinforcements. Onset cure temperature also varies depending on composition in the range of 120°C-240°C. The resulting polybenzoxazines have Tg's ranging from 220°C to 320°C. It was found that adding furan rings into the backbone of the polybenzoxazine network provides the added benefit of 800°C char yield as high as 70% as measured by TGA temperature ramps in an inert environment. In addition to promising processing and thermal characteristics useful for applications requiring high temperature ablation resistance, these materials contain high renewable content.

607251 Direct Production of Renewable Polyesters with High Glass Transition from Lignocellulosic Biomass

Abstract:

The need for sustainable polymers has driven the development of an array of chemical and biological pathways to potential monomers derived from renewable sources[1–5]. Despite decades of innovation, low conversion efficiencies from raw lignocellulosic feedstocks persist[6]. Here we present a new class of renewable polymer precursors that can be produced in a single step from the hemicellulose fraction of lignocellulosic biomass in high yields using a previously described aldehyde-facilitated biomass pretreatment[7–9]. By leveraging the use of aldehydes with different functionalities, we can install targeted functional handles onto both sides of the hemicellulose-derived xylose—creating a tricyclic, difunctional structure that presents as a potential renewable monomer for high-glass transition bioplastic synthesis. As the functional backbone of the aldehyde in our process can be varied, this strategy could unlock a new class of renewable polymers that can be produced in a single step from lignocellulosic biomass. These polymer precursors can also be produced directly produced from xylose, providing a direct route to high volume production of these bioplastics, circumventing the immediate need for a biorefinery. Here, we demonstrate the use of a carboxylic acid-functionalized aldehyde (glyoxylic acid) to produce a novel, tricyclic, diacid precursor in a solventless reaction directly from xylose up to 93% yield. Mechanically useful, high molecular-weight polyesters (50-180 kDa) with glass transition temperatures ranging from 76-134°C are then synthesized via melt condensation with commercially-relevant zinc and antimony-based catalysts (<100ppm) using a range of aliphatic diols. Chemical recycling of these polymers via low temperature hydrolysis and alcoholysis is demonstrated along with hydrolytic stability studies. As a proof-of-concept, the tricyclic, diacid monomer is produced directly from lignocellulosic biomass in concert with glyoxylic acid stabilized lignin and highly digestible, cellulose-rich solids. Finally, a techno-economic analysis generated using Aspen Plus is presented demonstrating the economical production of the dimethyl ester monomer from xylose.

References:

1. Zhu, Y., Romain, C. & Williams, C. K. Sustainable polymers from renewable resources. *Nature* **540**, 354–362 (2016).
2. Zhang, X., Fevre, M., Jones, G. O. & Waymouth, R. M. Catalysis as an Enabling Science for Sustainable Polymers. *Chem. Rev.* **118**, 839–885 (2018).
3. Vilela, C. *et al.* The quest for sustainable polyesters – insights into the future. *Polym. Chem.* **5**, 3119–3141 (2014).
4. Isikgor, F. H. & Becer, C. R. Lignocellulosic Biomass: A Sustainable Platform for Production of Bio-Based Chemicals and Polymers. 61.
5. Grignard, B., Gennen, S., Jérôme, C., Kleij, A. W. & Detrembleur, C. Advances in the use of CO_2 as a renewable feedstock for the synthesis of polymers. *Chem. Soc. Rev.* **48**, 4466–4514 (2019).
6. Hillmyer, M. A. The promise of plastics from plants. *Science* **358**, 868–870 (2017).
7. Shuai, L. *et al.* Formaldehyde stabilization facilitates lignin monomer production during biomass depolymerization. *Science* **354**, 329–333 (2016).
8. Questell-Santiago, Y. M., Zambrano-Varela, R., Talebi Amiri, M. & Luterbacher, J. S. Carbohydrate stabilization extends the kinetic limits of chemical polysaccharide depolymerization. *Nature Chemistry* (2018) doi:10.1038/s41557-018-0134-4.
9. Lan, W., Amiri, M. T., Hunston, C. M. & Luterbacher, J. S. Protection Group Effects During α,γ-Diol Lignin Stabilization Promote High-Selectivity Monomer Production. *Angewandte Chemie International Edition* **57**, 1356–1360 (2018).

599705 Optimization of Chitin Nanowhisker Deacetylation for Utilization in Renewable Oxygen Barrier Films and Coatings

Abstract:

The low oxygen permeabilities (OP) of cellulose and chitin nanocrystals have generated considerable interest in their application in food and beverage packaging as renewable substitutes for petroleum-based plastics. One approach is to apply aqueous suspensions of chitin and cellulose nanomaterials as coatings on a substrate film to improve its oxygen barrier property. Because chitin is cationic in acidic solution while sulfate-modified cellulose is anionic, the electrostatic attraction of the opposite charges may allow the fabrication of their nanomaterials into multilayered coatings. When such coatings dry very slowly, over several days, the nanofibers or nanocrystals pack densely and create low OP structures. However, when coating and drying are accelerated to realistic manufacturing throughputs, the OP of the resulting films increases and barrier performance is lost. This work demonstrates how the OP of high-throughput coated chitin and chitin-cellulose films can be improved considerably through optimization of the degree of acetylation (DA) of the chitin nanowhisker (ChNW). The DA is the percentage of acetamino groups to the sum of amino and acetamino groups, and indicates the charge density on the surface of ChNW.

In this work, ChNW and cellulose nanocrystals (CNC) were spray-coated on cellulose acetate (CA) film in an alternating sequence. The deacetylation of chitin extracted from crab shells was optimized to improve the oxygen barrier property of the coated film. Three factors influencing the deacetylation reaction were adjusted: concentration of sodium hydroxide (NaOH), reaction temperature and reaction time. The Taguchi method was applied for the design of experiments, the determination of the optimal deacetylation conditions and the relative effect of each factor.

The optimal conditions in the specified region of study were determined as 35 wt%, 140°C and 140 min, and the order of the contribution of each factor was temperature (68.0%) > NaOH concentration (21.7%) > time (10.3%). The ChNW prepared under more aggressive conditions resulted in better oxygen barrier properties (lower OP). In layered coatings containing ChNW and CNC on CA, the OP decreased from 1550 $cm^3 \cdot \mu m/m^2/day/kPa$ for uncoated CA, to 44.0 $cm^3 \cdot \mu m/m^2/day/kPa$ for coating without ChNW optimization, to a minimum of 16.7 $cm^3 \cdot \mu m/m^2/day/kPa$ for the optimal coating. This result is comparable to the OP of oriented poly(ethylene terephthalate). In addition, the coating led to ~20% improvement in water vapor transmission rate and similar mechanical properties to those of the neat CA film.

Furthermore, we present the preliminary results of improving the oxygen barrier property with the addition of chitosan to ChNW. Chitosan is highly deacetylated chitin, the DA of which is <10% in this work. When the mass ratio of chitosan to ChNW prepared under more aggressive conditions was 1:1, the OP of the coated CA film with ChNW+chitosan and CNC decreased to 8.9 $cm^3 \cdot \mu m/m^2/day/kPa$. This improvement showed the possibility of increasing the charge density to improve the oxygen barrier property.

601007 Reverse Dialysis to Concentrate Cellulose and Chitin Nanomaterials and Their Polymer Composite Gels

Abstract:

Cellulose nanomaterial is a biopolymer material from abundant cellulose sources, such as wood and plant. Two types of nanocellulose are mostly studied: cellulose nanocrystals (CNC), which consist of rod-like colloidal particles; and TEMPO oxidized cellulose nanofibrils (TEMPO-CNF), which are flexible fibers. Both CNC and TEMPO-CNF have negative surface charges. Chitin nanofiber, another biopolymer material, which is extracted from crustacean shells, are also flexible fibers, but with positive surface charge. All of these nanomaterials have high surface area-to-volume ratio, making them attractive to be used in gels or as reinforcing agents in a polymer matrix for applications like coatings, hydrogels, and composites. These applications require that the nanomaterial is fully dispersed to take advantage of its nanoscale morphology, which is difficult to achieve and is often only successful at low concentrations. In contrast, a high concentration of nanomaterial is often desired to improve the mechanical properties of gels and composites. Unfortunately, increasing the concentration by conventional methods like drying often results in aggregation. A processing method is needed to concentrate these nanomaterials and retain the benefit of large surface area.

In this work, we will show the potential of reverse dialysis as an effective and controllable method to concentrate cellulose and chitin nanomaterials and their PVA composite gels. Water is removed by dialyzing suspensions of the nanomaterials against a concentrated polymer solution. Although this method has been proposed before to concentrate nanocellulose, prior studies did not provide much detail. We first investigated the water removal kinetics at different PEG concentrations to understand the dialysis process and were able to concentrate CNC and TEMPO-CNF to at least twice the concentration of the existing commercially available slurry products. Rheological characterization of the samples dialyzed to various concentrations will be shown. Secondly, in addition to concentrating pure cellulose and chitin nanomaterials, we also use reverse dialysis to obtain high loadings of well-dispersed nanomaterials in polymer composite gels. The TEMPO-CNF/PVA composite gels were prepared in suspension at low concentration to ensure full dispersion, followed by water removal via reverse dialysis to form the gel. We will show that the composite gels prepared by this method are homogeneously dispersed at high loadings of polymers and nanomaterials, which cannot be easily achieved by directly mixing concentrated solutions of pure components. We used the new procedures to make PVA composite gels that contain TEMPO-CNF of different fibril sizes in order to study the effect of nanomaterials morphology. The reverse dialysis method enables quick processing of the materials, and samples can be retrieved with little loss. The method also increases the processing range of the cellulose and chitin nanomaterials, which can help to expand applications of these sustainable nanomaterials.

610572 Characterizing Network Structure in Novel Hydrogel Composites Containing Fractionated, Purified Lignins for Aqueous Separations

Abstract:

The fabrication of lignin-containing hydrogels has garnered attention for use in a variety of aqueous-based separations as lignin is a sustainable, naturally abundant biopolymer with a high concentration of hydroxyl groups, which can be utilized as crosslinking sites during hydrogel fabrication. However, one standing issue is our lack of understanding as to how the addition of lignin alters the network structure of these composite hydrogels given that the majority of lignin-based hydrogel studies utilize highly disperse, heterogeneous lignins, referred to as crude bulk lignins (CBLs). Herein, a novel series of lignin–poly(vinyl alcohol) (PVA) composites were synthesized utilizing ultraclean lignins (UCLs) of prescribed molecular weights (MWs) and low dispersity using two different crosslinking agents (CLAs) – methylenebisacrylamide (MBA) and glutaraldehyde. Prior to hydrogel formation, lignin hydroxyl groups were functionalized with vinyl groups, which was confirmed via ^{1}H and ^{31}P nuclear magnetic resonance spectroscopy, allowing the lignins to chemically crosslink with PVA chains. The permeability of various pollutants (e.g., methylene blue) through the hydrated composites was measured via ultraviolet-visible spectroscopy, where penetrant permeability was found to depend on the MW of both the lignins and PVA, the source of the lignin prior to fractionation, as well as the concentration of CLAs utilized during membrane fabrication. In addition, poroelastic relaxation indentation (PRI) was used to characterize both the mechanical and transport properties of the composites. Namely, water diffusivity, hydrated mesh size, and Young's modulus of the composites were obtained from PRI measurements. Results from this work indicate that transport of pollutants through the composite hydrogels is governed by a combination of the network structure and the strength of interaction between the pollutant and lignin.

607325 Ion Conducting Polymers from Kraft Lignin for Electrochemical Devices

Abstract:

An innovative lignin valorization effort, designing lignin-rich waste as efficient and cost-effective energy materials, can aid in both bio- and energy economy. One major technical limitation of electrochemical devices (e.g., fuel cells) is the weak ion conductivity within the 2-30nm thick catalyst binding, ion-conducting polymer (ionomer) layer over electrodes. Also low-cost ionomer formulations can lower down the production cost of fuel cells. In this work, we strategically sulfonated kraft lignin (a by-product of pulp and paper industries) to design ionomers with varied ion exchange capacities (IECs) (LS x; x=IEC) that is cheap, environment friendly and can potentially overcome this interfacial ion conduction limitation. Unlike commercial lignosulfonate, the water solubility of LS x was overcome by controlling the sulfomethylation and cross-linking reactions. The proton conductivity, water uptake, ionic domain characteristics, density, and water mobility were measured in sub-micron thick LS 1.6 and LS 3.1 films and compared with the current state-of-the-art ionomer, Nafion thin films. LS 1.6 showed much higher ion conductivity than Nafion and LS 3.1 in films with similar thickness despite of their water uptakes. Within the three-dimensional, less dense, branched architecture of LS 1.6 macromolecules, the $-SO_3H$ and $-OH$ groups are in close proximity facilitating the formation of larger ionic domains with highly mobile water molecules. As compared to LS 1.6, LS 3.1 showed a higher glass transition temperature and film stiffness at dry state, which did not change upon humidification. On the other hand, Nafion stiffened significantly upon humidification only. These results demonstrate the potential of LS x as an ideal candidate as ionomer binder for low-temperature, water-mediated ion conduction in energy conversion and storage devices.

593129 Advanced Fabrication and MULTI-Properties of Rubber Aerogels from CAR Tire Waste

Abstract:

End-of-life tires should be considered as one of the fastest-growing waste streams in the world. The annual production of waste tires is estimated to amount to 1.5 billion whole-tires worldwide. This study, for the first time, develops a simple but cost-effective method to fabricate a novel material- rubber aerogel using recycled tire polymer fibers (RTPF) shredded from car tire waste, polyvinyl alcohol (PVA) and glutaraldehyde (GA) by freeze-drying method. They are extremely light, highly absorbent, very durable, and they are also very efficient at trapping heat and sound. Rubber aerogel has ultra-low density (0.020–0.091 g/cm^3), high porosity (92.2–98.3%) and excellent heat insulation properties (0.035 – 0.049 W/m.K). It has a high noise reduction coefficient (0.56) and performs 1.4 time % better than commercial sound foam absorber. The oil absorption capacity of the rubber aerogel is measured up to 25 g/g, approximately twice the absorption capacity of conventional absorbent like polypropylene mat. Rubber aerogel also shows very good durability as it springs back to its original shape after compression tests. By upcycling waste rubber into products of a higher commercial value, this new technology promotes the wider use of scrap tyres and offers an eco-friendly way to recycle used rubber.

610107 Exploring Degradation As a Viable End-of-Life Process for Thermoset Polymers

Abstract:

There is a great need to recycle polymers to minimize their environmental impact, as the majority of produced polymers end up in landfills after their useful lifetime. Linear polyesters degrade rapidly, and their degradation behavior has been well studied, which showed surface erosion in basic media and bulk erosion in acidic media. By contrast, thermoset polymers normally require harsh degradation conditions due to their highly crosslinked structures. We have explored the incorporation of cleavable hydrolytic linkages into the crosslinked network, to determine if thermoset polymers can be also degraded under benign conditions. Epoxy resins are thermoset polymers widely used in composites, coatings and adhesives, with applications spanning automotive and aerospace industries, structural components, and wind turbine blades, among others. The standard precursor to epoxy resins, the diglycidyl ether of bisphenol A (DGEBA), is derived from petroleum and lacks functional groups which can promote degradation under benign conditions. Epoxidized vegetable oils, phenolic acids, and vanillic acid, which contain ester linkages, were investigated as sustainable sources to produce degradable epoxy resins. The resulting phenolic acid-based and vanillic acid-based epoxy resins exhibited comparable thermal and mechanical properties to conventional DGEBA-based epoxy resins. The accelerated hydrolytic degradation behavior of the ester-containing epoxy resins was explored, through monitoring of the polymer mass loss after exposure to a basic solution at moderate temperatures. The biobased epoxy resins exhibited rapid degradation in the basic solution, in contrast to the slow degradation rate of the traditional DGEBA-based epoxy resin. Various factors which affected the degradation rates were discussed. A degradation model was proposed to describe the degradation behavior and mass spectrometry confirmed the mechanism was surface erosion through ester hydrolysis in basic media. For comparison purposes, accelerated hydrolytic degradation behavior under mild acidic conditions was also explored. The data showed biobased epoxy resins exhibited bulk erosion in acidic media, in agreement with the behavior of linear polyesters. Another degradation model was proposed to describe the degradation behavior in acidic media and mass spectrometry and Fourier-transform infrared spectroscopy confirmed the mechanism was bulk erosion through ester hydrolysis.

614005 Finding Our Way to Drive Sustainability Forward Through a Global Pandemic

Abstract:

Ford commitment to sustainability does not disappear during a pandemic, resource scarcity, low oil price, or with many other challenges. The presentation will outline how the growth of the automotive industry and R & D efforts encourage the use of cost effective, lightweight, sustainable and advanced materials for automotive applications. Pioneering developments of sustainable and advanced materials by Ford scientists including nanocomposites, aerogels, natural fiber reinforced composites, polymeric and soft materials made from renewable feed stocks, 3-D printed plastic parts, bio-inspired and patterned functional materials as well as plastic parts made from recycled carbon dioxide will be featured.

614006 Creating New Bioproducts Targeting "Zero Waste" Production

Abstract:

This presentation outlines USDA strategies for converting agricultural coproducts into commercially-viable products with the ultimate goal of achieving "zero waste" agriculture. The Western U.S., especially the Central Valley of California, is a world leader in the production of an array of specialty crops, including wine grapes, olives, artichokes, avocados, almonds, walnuts, pistachios, and, most recently, hemp. These crops contribute billions of dollars to the rural economy. The processing of each of these crops generates their own specific coproducts and waste streams, which in turn can generate billions more in revenue or result in significant costs. This presentation will discuss successful development of several ag-derived biomass feedstocks into commercial products, outlining specific processing methods that meet the cost-points required to meet market demands. Discussion will include description of torrefaction of biomass whereby biomass sources were heat-treatment to 200 C-300 C in limited oxygen, to create fillers for modifying the properties of traditional plastics. Char-like ground torrefied additives were melt-blended into recycled plastics and bioplastics as a replacement for carbon black. In another example, we describe converting organic acids and/or methane from agricultural biomass into polyhydroxyalkanoate (PHA), a commodity bioplastic with significant market potential. As will be discussed, the drive toward "zero waste" agriculture can be applied universally to generate value-added bioproducts from a wide array of agriculturally-derived biomass sources.

614007 Can Bioplastics Alleviate Plastic Waste Issues?

Abstract:

Plastic wastes on land and in oceans have become major societal issues. Articles in print, television, and social media about plastics waste issues and bans on plastic items are on the rise everywhere in the world. Most serious, is the issue relating to microplastics found in oceans and living systems. Against this backdrop, this lecture reviews the role of BioPlastics in providing solutions for plastic waste issues. There are many misperceptions about bioplastics, biobased, biodegradable, compostable, and the value proposition they offer for addressing the plastic waste issues. In this context, this @EnvSciTech viewpoint summarizes necessary requirements for assessing and reporting plastic biodegradation as solution for plastic waste https://pubs.acs.org/doi/10.1021/acs.est.9b04513. The presentation reviews the science, the definitions, standards, and misleading claims that abound in the marketplace.

614008 Circular Economy and Sustainable Materials

Abstract:

Plastic waste and their adverse environmental impacts have become daily front news. Since 1950, we have accumulated more than 7.5 billion tons of plastic wastes in the globe. These wastes quite often end-up in in ocean. According to United Nations, each year we create around 300 million metric ton of plastics wastes that is equivalent to the weight of our entire human population. Similarly, billion tons of agro-residues and forestry wastes are unused or burnt very often-creating environmental pollution.

Agro-residues and forest wastes including waste wood and sawdust on reinforcement with waste plastics can create a range of sustainable hybrid biocomposites. Valorization of waste plastics in biocomposites uses is a strategy towards lessening green house gas emissions. Giving new birth to waste plastics supports circular economy and this is a strategy towards improved sustainability. The sustainable biocomposites can find applications in consumer products, rigid packaging, and green housing structures as well as lightweight automotive parts.

595838 Hydrothermal Liquefaction for Co-Valorization of Plastics and Biopolymers

Abstract:

Hydrothermal liquefaction (HTL) takes place in water at near- and supercritical conditions (T_c = 374 °C, P_c = 22 MPa). HTL has been studied to valorize components of municipal solid waste (MSW), namely synthetic and biopolymers. Biopolymers such as those in food waste and wood are broken down to produce energy-dense oil under subcritical conditions (280-374°C) while synthetic polymers generally decompose under supercritical conditions (374-500°C). HTL of plastics (polypropylene (PPE), polyethylene terephthalate (PET), polystyrene (PS), polycarbonate (PBC)) can give higher yields of energy dense (33 – 91 wt% oil yield) than HTL of biopolymers (10 – 50 %). The heating value of the oil produced from some plastics is higher than that from biopolymers and comparable to petroleum. The oil, solid, and aqueous products obtained contain moieties similar to monomeric units of plastics that could be recycled.

Separation of biopolymers from plastics for recycling has been a challenge due to contamination, source management and enormity of waste produced. This limitation is extended to most chemical and mechanical recycling techniques. HTL offers a unique platform to valorize together most components of municipal solid waste to obtain useful products while bypassing the need for separation. The products obtained are separated into 4 phases – non-polar liquid, polar liquid, solid, and gas. HTL can be hence be viewed as an intensified step that combines reaction and separation.

Co-liquefaction experiments were performed with mixtures of nine model compounds (5 biopolymers, 4 plastics) to mimic the composition of MSW. Experiments were conducted from 300 – 425 °C with a batch holding time of 30 min. Synergistic interactions were calculated by comparing the oil yield of the mixture with the oil yields from HTL of the biopolymer and plastics individually. Note that synergistic interactions are observed among plastic, biopolymers in their respective mixtures. The percentage synergy in the nine-component mixture exceeds the synergies obtained among the separate mixtures of plastic alone and biopolymer alone. This indicates the presence of positive interactions between plastics and biopolymers that increase the oil yield and hence the energy recovered from the process. A high energy recovery of 45 percent was obtained at the lowest operated temperature (300 °C). The largest percentage synergy (105 %) is also observed at the same temperature. The presence of these synergistic behaviors at a low operating temperature aids at obtaining higher energy recoveries at lower operating temperatures which is economically and environmentally beneficial.

To further pinpoint the binary interactions between biopolymers and plastics that cause such synergy – more experiments were conducted by manipulating the mixture composition. We will also report on a mathematical model that would predict the energy recovered from isothermal operation of HTL for any given mixture of synthetic polymers and biopolymers. This would further aid in enabling feedstock engineering to increase the net energy recovered to more than 45% i.e. adding waste feedstocks to optimize yield obtained from available waste feedstock.

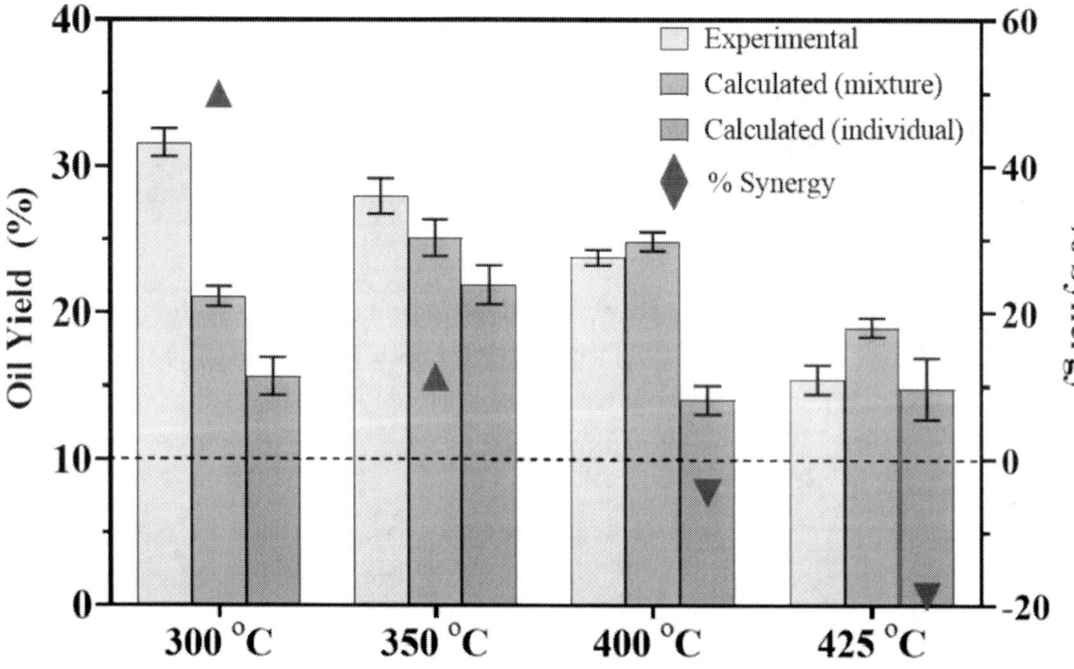

Fig: Oil yield from HTL treatment of MSW mixture based on model compounds (cellulose, starch, lignin, soy protein and lipid, PPE, PBC, PS, PET) at different temperatures with calculated (individual) weight averaged yield based on experiments on model compounds, calculated (mixture) weight averaged yield based on experiments on plastic and biopolymer mixture and % synergy

613277 Polystyrene Sulfonic Acid As a Recoverable Catalyst for the Hydrolysis of PET

Abstract:

Poly (4-styrenesulfonic acid) (PSSA) was applied as a recoverable acid catalyst to hydrolyze polyethylene terephthalate (PET) to terephthalic acid (TPA) and ethylene glycol. The impact of proton concentration and temperature on depolymerization of PET were compared for PSSA and H_2SO_4. The PSSA system exhibited more rapid hydrolysis kinetics than the sulfuric acid at all conditions studied in this work. The hydrophobic backbone of PSSA improved surface wetting of the PET as measured by contact angle. The increase in reaction rate was attributed to higher localized concentration of protons at the surface of the PET by the PSSA solutions. The PSSA maintained catalytic activity after use and recovery for five cycles at 150 °C.

613276 Conversion of Waste Solids to Monomers and Intermediates

Abstract:

In this study a process design and assessment for a waste plastics recycling plant to produce monomers and refinery-like liquids for use as intermediate chemicals is presented. The catalyst is comprised of a promoted and modified ZSM-5 zeolite catalyst spray dried into an attrition resistant, fluidizable, spherical shape to be used in a circulating fluid bed reactor. A feed stock of mixed, co-mingled solid waste composed of plastics, polyols, polyethers, polyurethanes, paper, rubber, tires, auto shredder residue, composites, agricultural waste, MRF bales (including rejected and residuals) and the like is added to a fluidized bed reactor in a melted state and the fluidized catalyst bed is operated in the range of 550 – 600° C at 1.5 – 3 atm, typically. The catalyst regeneration will take place at 600 – 750° C. The mass balances are in the range of 100 $\pm$ 10% and the analytical measurements of the reaction products are conducted by a duel function gas chromatograph equipped with a flame ionization detector and thermal conductivity detector. The light products wt% yields, CO, CO_2, H_2, C_1-C_3, sum to below 20 wt% yield and the wt% liquids falls above 80%.

601264 Valorization of Plastic Waste Using Hot Pressurized Water

Abstract:

Plastic waste can be processed in hot pressurized water (hydrothermal liquefaction) to recover energy (oil) and chemicals (monomers). Most waste plastics (76 wt%) are landfilled, with just 9 wt% being mechanically recycled and 15 wt% being burned for energy recovery. The need for separation of plastics, costs, and environmental impact has greatly curbed the growth of mechanical recycling. Direct energy recovery processes such as combustion release pollutants such as oxides of sulfur and nitrogen. Pyrolysis is a potentially viable process for valorizing some plastics (e.g., polyolefins), but it is not well suited for others (e.g., polyesters). If done at scale, pyrolysis will likely require feedstock separation, both of different types of plastics and of plastics from non-plastics. This difficulty might be overcome by processing plastics in hot pressurized water.

Hydrothermal liquefaction (HTL) takes place in water at near- and supercritical conditions (T_c = 374 °C). Water can act as a solvent, reactant, or catalyst for an array of acid- or base-catalyzed organic chemical reactions. These favorable conditions aid in depolymerization reactions at relatively low temperatures. In this study, we focused on the energy recovery in the oil phase and monomer recovery in oil, aqueous, and solid phases from HTL of polyesters - polyethylene terephthalate (PET), polycarbonates - poly bisphenol-A carbonate (PBC) and polyolefins - polypropylene (PPE), polystyrene (PS) at 300 – 450 °C and 30 – 60 mins batch holding times.

Energy-dense oils are produced from not just polyolefins but polycarbonates and polyester as well. PPE and PET produce considerable oil under supercritical conditions (425 °C), while PS and PBC produce considerable oil even under subcritical conditions (350 °C). Plastics other than polystyrene gave the highest energy recoveries at 425 °C, 30 mins – PPE (29 %), PBC (66 %), PET (35 %). This means that different plastics can be processed using HTL under the same process conditions to achieve optimal energy recoveries. Hence, HTL can recover energy efficiently in the oil phase from plastics without the need for any separation.

The water in the HTL process acts as a reactant and several products obtained from PBC and PET (as detected from GC/MS) likely arose from hydrolysis. Also, PPE (a hydrocarbon) mainly produces long-chain alcohols. Hence, water molecules interact with PPE or PPE-derived intermediates as the PPE decomposes. Correlations between GC/MS, oil yield, and CHNS analysis show that dehydration reactions (occurring at 400 – 425 °C) cause an increase in heating values in oils from HTL of PBC and PET, in turn, resulting in high energy recoveries at these temperatures. At higher temperatures (450 °C), aromatization becomes the major pathway and leads to a reduction in heating value (oil phase).

The suitability of HTL for chemical recycling of plastics depends on the individual plastic in view. We found high concentrations of Bisphenol A in the oil phase from HTL of PBC at 300 °C. Note that no other plastic decomposes at 300 °C. Hence, differential HTL processes can be used to recover selective monomers from a mixture of plastics. PET is another material from which recovery of chemicals is possible. Benzoic acid (derivative of terephthalic acid) was the major product in the oil phase from HTL of PET at higher temperatures (450 °C). Thus, there may be opportunities to recover chemicals, along with oil, from the product streams from HTL of plastics. We will also report on HTL processing of nylon 6, nylon 6,6, polylactic acid, and polysulfone under similar conditions.

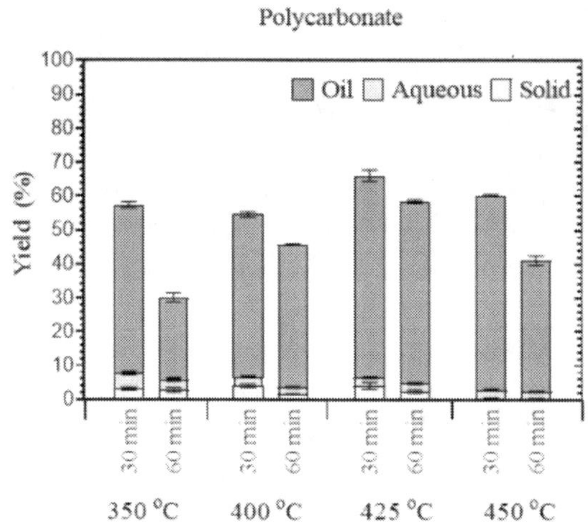

Top 3 compounds for PBC at various temperatures as detected from GC-MS									
300 °C		350 °C		400 °C		425 °C		450 °C	
Bisphenol-A	72	Benzene, 1-methoxy-2,4-dimethyl-	40	Benzene, 1-methoxy-2,4-dimethyl-	39	Benzene, 1-methoxy-2,4-dimethyl-	56	Benzene, 1-methoxy-2,4-dimethyl-	48
Phenol, 4-(1-methyl-1-phenylethyl)-	21	Phenol	17	2-Hepten-4-yne, 2-methyl-	19	Phenol, 4-(1,1-dimethylpropyl)-	13	Phenol, 3,4-dimethyl-	11
Phenol	2	Phenol, 4-(1,1-dimethylpropyl)-	15	Phenol	11	Bisphenol-A	12	Phenol, 4-(1-methylpropyl)-	6

Fig: Variation of product fraction yields (wt.%) from HTL of polycarbonate with temperature and time and top 3 concentrated products detected by GC/MS with corresponding percentage peak area at different temperatures with batch time of 60 mins.

AUTHOR INDEX

AIChE
120 Wall Street, FL 23
New York, NY 10005-4020

ISBN 978-1-7138-2295-0

CURRAN ASSOCIATES INC.
proceedings.com